AIDS TO BIRD IDENTIFICATION
IN SOUTHERN AFRICA

To Cherie

Aids
to Bird Identification
in Southern Africa

Second Edition

Gordon L. Maclean

Illustrated by
Linda Davis

UNIVERSITY OF NATAL PRESS
PIETERMARITZBURG
1988

First published 1981. Reprinted 1982. Second edition 1987
ISBN 0 86980 261 5 (First edition)
ISBN 0 86980 586 X (Second edition)

Typeset in the University of Natal Press
Printed by The Natal Witness (Pty) Ltd, Pietermaritzburg

Contents

Preface   *vii*

Introduction   *ix*

    CHAPTERS

1   Names and classification of birds   1

2   Finding your way around   2

3   The waterbirds   10

4   Slender-billed arboreal passerines   11

5   The near-passerines   13

6   Five tricky groups   14

7   The raptors, or diurnal birds of prey   18

8   A word of encouragement   21

    KEYS

    Waterbirds   22

    Scolopacid waders   24

    Groups of slender-billed arboreal passerines   28

    Groups of near-passerines   30

    Woodpeckers   32

    Swallows   35

    Francolins   37

    Diurnal birds of prey (raptors)   39

    Larks   48

    Cisticolas   51

    Female and eclipse male widows, bishops and queleas   54

    Some common species-pairs and species-groups   56

Useful books on southern African birds   62

Preface

Since its appearance in 1981 this little book has met with such a favourable reception that it has been found necessary to update, expand and reprint it in a second edition. Birdwatching has become more popular than ever as a recreational activity and the output of bird books has not lessened in any way – field guides, checklists, coffee-table books –, to the extent that the beginner is bewildered by the array and often doesn't know what book to buy. No handbook yet produced in southern Africa can compare with *Roberts' birds of southern Africa* as both an identification manual and a source of ornithological information, although other fieldguides have appeared in the meantime; the best of these is Ken Newman's *Birds of southern Africa* (1983, Macmillan). *Roberts'*, however, is more than just a fieldguide and anyone even remotely interested in birds should have a copy. At the time of going to press, the latest *Roberts'* is the 5th edition (published in 1985 and reprinted in 1987) revised by myself, on which this set of *Aids* is based and which in fact includes much of the material, including the keys, that first appeared in the *Aids*.

Roberts' itself is a rather daunting book because there are so many birds illustrated that the beginner is discouraged right from the start. So, several years ago, when I was asked to give a course in bird identification to an adult-education class, I decided that the best way to start was to narrow down the field before wading through endless pages of pictures or text. And that is just what I set out to do.

The success of this course was immediate and since then I have given it many times throughout southern Africa. The pleasure these lectures have given my audiences and myself prompted many people to urge me to publish my handout in book form. The first person to do so was my wife, Cherie; and she is still my main source of encouragement. To her this book is dedicated and to her goes my warmest gratitude for her help, for her constructive criticism, and for attending many of the courses I have given.

This book is adapted more or less directly from my lecture courses. I have deliberately retained the lecture idiom so as to preserve the informal relationship that I have established with students. I hope this makes the book easier to read and to learn from. Although it is aimed at the absolute beginner, many experienced birdwatchers have attended my courses and claim to have gained considerably from them.

At times my *Aids to bird identification* has been likened to 'a set of mathematical puzzles' and I have to agree that this is so, at least until you get used to the approach. The idea is to remove the hit-and-miss kind of identification attempts usually made by beginners, and to introduce a measure of precision into both the process and the result. This may at first seem to take the fun out of birding, but this is the last thing I'd want to do. Birding is tremendous fun, and especially so when even your earliest attempts at identification are crowned with success. And that is what I hope this book will help you to achieve.

The University of Natal Press and I are grateful to the Trustees of the John Voelcker Bird Book Fund for allowing us to use material from the 1985 *Roberts' birds of southern Africa* in this edition of *Aids*; this includes the keys to the scolopacid waders and to the woodpeckers. I have also prepared two new keys for this edition of *Aids*, those to the cisticolas and to the bishops, widows and queleas.

Introduction

This set of *Aids to bird identification in southern Africa* is intended to assist the most helpless beginner to identify almost any bird south of the Zambezi and Cunene rivers.
The equipment is simple. All you need is:

(a) a copy of *Roberts' birds of southern Africa* (1985 edition), or any other field guide;
(b) a pair of binoculars — preferably 7 x 35, 8 x 30, 10 x 40 or some similar size;
(c) a small field notebook, about 10 cm x 7,5 cm in size;
(d) a pen or pencil.

Later on you may decide to extend your reading, so on page 62 you will find a list of relevant bird books to help you with your birding and to give you more information about the birds themselves.

The idea behind these *Aids* is to help you to know where to start to look for your bird in *Roberts*', so that you don't have to work through more than 800 pages of text and over 1 600 illustrated birds each time you see something unfamiliar. The book is designed to fit into your copy of *Roberts*' or into a sling bag designed to take both books into the field for use on the spot. A lot of people, including myself, keep a copy of *Roberts*' for field use and a second copy for library use. The *Aids* are largely based on the principle of the dichotomous key — that is a key with two choices at a starting point (always Number 1) and two choices at each number thereafter until your bird, or at least the group to which it belongs, is 'keyed out'. Some of the numbers involve three choices instead of two; this is just an economy measure to save a step in the keying out, but is doesn't make things any more difficult.

Before referring to the keys get a good look at your bird (with your binoculars). Write down every feature that you think may be useful, and make a simple sketch too, if you can (in your small field notebook); note down also the habitat and then decide on which key is the correct one to use. Then start at Number 1 at the top left margin, make a choice and run your eye along the line to the right-hand margin where you'll find a number. That number refers to your next choice. Find the number in the left-hand margin again, make your next choice and repeat the process until you arrive at an answer (the correct one you hope!). The numbers of pages and Plates where you will find text and illustration of your bird have been added to the keys to speed things up for you when checking against the key. **The Plate numbers are in bold type in parentheses after the name of the bird or group.** The method will become much clearer when you actually begin to use a key, but there is a bit of homework to do before that. So let's start.

Chapter One

Names and Classification of Birds

All living things are classified according to their relationships, or degrees of likeness to one another. Birds are characterized by having feathers and having the forelimbs modified to form wings; all birds share these likenesses.

The Class **Aves**, as the birds are known, is subdivided into Orders, Families, Genera (sing. Genus) and Species. The smallest relevant subdivision is the species: all members of the same species look alike, although sex or age differences may exist.

Thus all Glossy Starlings look alike, and belong to the genus *Lamprotornis*, species *nitens*. It is usual to say the specific name is *Lamprotornis nitens*. The Glossy Starling may be mistaken by the beginner for one of the other 'glossy' starlings because they are similar though by no means identical. Because of their similarities the 'glossy' starlings all belong to the genus *Lamprotornis*; thus we get

Lamprotornis corruscus . . . Blackbellied Starling
Lamprotornis chalybaeus . . . Greater Blue-eared Starling

and so on. Other starlings belong to other genera, but all starlings belong to the single family Sturnidae, the starling family.

The idea of families, genera and species is relatively easy to grasp. Divisions (or groupings) higher than this, however, are less easy to grasp and also less easy to recognize. The roughly 160–170 families of living birds are grouped together into 28–30 orders. For example, six families of waterbirds with all four toes webbed together (a condition known as totipalmate) form the order Pelecaniformes: these families are the tropicbirds, pelicans, boobies and gannets, cormorants, darters and frigatebirds.

Most orders of birds consist of fewer than 20 families, but one order, the Passeriformes, comprises over 60 families and contains more than half of all living bird species. The Passeriformes, or 'passerines' as they are more familiarly known, are mostly smallish birds (the crows are exceptional) characterized by a foot adapted to perching by having three toes in front and one behind. The toes are seldom joined together (the broadbills are exceptions here), cannot be placed in any other position, and usually cannot be used for catching or grasping prey. It is difficult to give a more definite and obvious set of features for the passerines, but one develops a 'feel' for them as one becomes more familiar with them. In *Roberts'* the passerines can be found from page 424 to 792. Try to familiarize yourself with the kinds of birds included in this order.

All the other orders of birds are simply lumped together by the rather artificial term 'non-passerine'. This doesn't really tell one much until one is familiar with who the passerines are, so it's a worthwhile exercise.

Among the non-passerines are several orders of arboreal (tree-dwelling) birds which could be mistaken for passerines by a beginner. It took me years before I really worked out who the passerines were, so take heart! But these confusing non-passerines that I have mentioned can conveniently be called 'near-passerines'. They form a sort of shadow group between birds that are obviously passerine and those that are obviously non-passerine. Only experience will help one to sort out the members of these three groups: passerines, near-passerines and non-passerines.

Chapter Two

Finding Your Way Around

The first thing to find your way around is your fieldguide – in this case *Roberts' birds of southern Africa*. Keep it in your car, at your bedside or next to your favourite chair – wherever you'll be able to pick it up and browse through it in a leisurely moment. Study its layout. You'll soon find that each species is treated in a similar way.

Firstly there is the new National Bird Number in bold type, followed by the old '*Roberts*' number in parentheses. In the centre are the English and Afrikaans names and on the right is the Plate number, also printed bold. Under the bold names are the scientific names in *italics*. At the start of most of the species accounts is a set of African names in small type. Then follows the rest of the text, divided by bold subheadings into: Measurements, Bare Parts, Identification, Voice, Distribution, Status, Habitat, Habits, Food and Breeding. The method of treatment of each of these sections is treated in detail in the Introduction to *Roberts'*, the set of pages known as the 'prelims', preceding the colour plates and numbered in small roman numerals.

In the 1985 *Roberts'* the section on subspecies (or 'Local Races') has been done away with to save space. Subspecies are geographically variable forms of the same species, sometimes quite markedly different in coloration from one part of the country to the next. For example Fiscal Shrikes in eastern South Africa have the top of the head all black; those in the west have a broad white eyebrow (Figure 1). Where subspecies are recognizably different in the field, the extreme forms have usually been illustrated in the colour plates or discussed in the text.

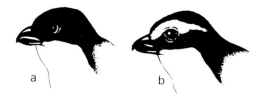

Figures 1 **Heads of two geographical races of the Fiscal Shrike.**
a = eastern form with no white eyebrow (*collaris*).
b = western form with broad white eyebrow (*subcoronatus*).

Accompanying nearly every species account is a small map of southern Africa with part of the area shaded in colour to show where the species normally occurs. The small scale of these distribution maps means that they are a rough guide only, and not to be taken as precise and invariable indications of where a bird can be found.

Let's go back to the measurements at the start of each account. The 'Length' of the bird may seem deceptive or exaggerated at first. This is because the measurement is taken in a standard way from museum specimens placed on their backs and measured straight out from tip of bill to tip of tail. But by choosing a well known bird, as I have suggested, and comparing its length in the book with that of its actual size in life, you will get an idea of what size category you may be seeing in the field.

Now it's time to do some real homework. The idea of the following exercise is to familiarize you with *Roberts'*, with the names of birds, and with the kinds of birds you are likely to see in your area. Decide on a place where you'd like to go (or have already been) birdwatching. On a sheet of paper make a list of the kinds of habitat you'll find in the area you have chosen, keeping the area small enough so that you have no more than six or seven habitats. (A *habitat* is simply a recognizable type of countryside.) Let's take typical example from the eastern part of the country. My own area aro..nd Pietermaritzburg will have a habitat list something like this:

1. Urban (city area)
2. Suburban
3. Grassveld
4. Thornveld
5. Forest
6. Freshwater (dam, river, etc.)

A list of habitats from the arid western parts of the southern African region (say around Keetmanshoop in Namibia/SWA) might look something like this:

1. Plains (scrubby or grassy)
2. Rocky hills
3. Trees along dry watercourse
4. Town
5. Farmyard
6. Dam

Now take another sheet of paper and make a broad left hand column for the bird species, and as many more vertical columns as you have habitats in your list, thus:

Species	Habitat				
	Urban	Suburban	Grassveld	Thornveld	etc.

Now take your copy of *Roberts'* and, starting at page 1, check which species of birds are supposed to occur in your area according to the colour on the marginal distribution maps. Write down in the 'Species' column those that should occur, and check the habitat of each in the next and mark it in the relevant habitat column with a plus (+). If we take Pietermaritzburg as our example, the list will look like this:

Species	Habitat					
	Urban	Suburban	Grass- veld	Thorn- veld	Forest	Fresh water
6 Dabchick						+
47 Whitebreasted Cormorant						+
50 Reed Cormorant						+
52 Darter etc.						+

As you proceed, you'll notice an interesting thing: the groups of birds you deal with tend to fall into definite habitats. Most of the birds at the start of *Roberts'* are waterbirds, most of the francolins and other gamebirds are grassveld dwellers, most of the smaller birds near the end of *Roberts'* are arboreal (living in trees), and so on.

So let's make another list. This time I've done it for you in three separate lists: Non-passerines, Near-passerines and Passerines (Table 1). Let's have a close look at these lists. Each bird family is associated with a major habitat. Nearly all the non-passerines are aquatic, whether marine or freshwater (exceptions: the Ostrich, some birds of prey, gamebirds like francolins and korhaans, coursers, pratincoles and sandgrouse). Nearly all the near-passerines are arboreal (exceptions: some owls and the swifts). Most of the passerines are arboreal too. This can be summed up by saying that systematic groups of birds tend also to fall into habitat groups. And this makes life very much simpler as we shall see.

Table 1:   Southern African Bird Families and their Main Habitats

A   THE NON-PASSERINES

Family	Habitat
Ostrich (Struthionidae)	ground
Penguins (Spheniscidae)	aquatic (marine)
Grebes (Podicipedidae)	aquatic (freshwater)
Albatrosses (Diomedeidae)	aquatic (marine)
Petrels (Procellariidae, Oceanitidae)	aquatic (marine)
Tropicbirds (Phaethontidae)	aquatic (marine)
Pelicans (Pelecanidae)	aquatic (marine and freshwater)
Gannets and **Boobies** (Sulidae)	aquatic (marine)
Cormorants (Phalacrocoracidae)	aquatic (marine and freshwater)
Darter (Anhingidae)	aquatic (freshwater)
Frigatebirds (Fregatidae)	aquatic (marine)
Herons, Egrets, Bitterns (Ardeidae)	aquatic (freshwater)
Hamerkop (Scopidae)	aquatic (freshwater)
Storks (Ciconiidae)	aquatic (freshwater), ground
Ibises and **Spoonbill** (Plataleidae)	aquatic (freshwater), ground
Flamingoes (Phoenicopteridae)	aquatic (marine and freshwater)
Ducks and **Geese** (Anatidae)	aquatic (freshwater)
Secretarybird (Sagittariidae)	ground
*Falcons and **Kestrels** (Falconidae)	arboreal, cliff

Family	Habitat
*Vultures, Eagles, Buzzards, Harriers, Hawks, Kites, (Accipitridae)	arboreal, cliff
Osprey (Pandionidae)	arboreal, aquatic
Francolins and Quail (Phasianidae)	ground
Guineafowl (Numididae)	ground
Buttonquails (Turnicidae)	ground
Rails, Crakes, Moorhen, Coot (Rallidae)	aquatic (freshwater)
Finfoot (Heliornithidae)	aquatic (freshwater)
Cranes (Gruidae)	aquatic (freshwater), ground
Bustards and Korhaans (Otididae)	ground
†Jacanas (Jacanidae)	aquatic (freshwater)
†Painted Snipe (Rostratulidae)	aquatic (freshwater)
†Oystercatchers (Haematopodidae)	aquatic (marine)
†Plovers (Charadriidae)	aquatic (marine and freshwater), ground
†Snipe, Sandpipers, Phalaropes, etc. (= 'waders') (Scolopacidae)	aquatic (marine and freshwater)
†Avocet and Stilt (Recurvirostridae)	aquatic (marine and freshwater)
†Crab Plover (Dromadidae)	aquatic (marine)
†Dikkops (Burhinidae)	aquatic (marine and freshwater), ground
†Coursers and Pratincoles (Glareolidae)	ground
Skuas (Stercorariidae)	aquatic (marine)
Gulls (Laridae)	aquatic (marine and freshwater)
Terns (Sternidae)	aquatic (marine and freshwater)
Skimmer (Rynchopidae)	aquatic (freshwater)
Sandgrouse (Pteroclidae)	ground
Doves and Pigeons (Columbidae)	arboreal, ground

* Diurnal raptors (day-hunting birds of prey).
† These 9 familes form the group known collectively as the 'WADERS', as distinct from other long-legged waterbirds which are called 'wading birds'. The term 'waders' may also be applied only to the family Scolopacidae, the snipe, sandpipers and their relatives, which are identified in the Key to the Scolopacid Waders on page 24.

Family	Habitat
Parrots (Psittacidae)	arboreal
Louries (Musophagidae)	arboreal
Cuckoos and **Coucals** (Cuculidae)	arboreal
Barn and **Grass Owls** (Tytonidae)	ground, cave, building
Owls (Strigidae)	arboreal, cliff
Nightjars (Caprimulgidae)	ground, arboreal
Swifts (Apodidae)	aerial, cliff, building
Mousebirds (Coliidae)	arboreal
Trogon (Trogonidae)	arboreal
Kingfishers (Alcedinidae)	arboreal, aquatic
Bee-eaters (Meropidae)	arboreal, river bank
Rollers (Coraciidae)	arboreal
Hoopoe (Upupidae)	ground, arboreal
Woodhoopoes (Phoeniculidae)	arboreal
Hornbills (Bucerotidae)	arboreal, ground
Barbets (Capitonidae)	arboreal
Honeyguides (Indicatoridae)	arboreal
Woodpeckers (Picidae)	arboreal, ground
Wryneck (Jyngidae)	arboreal

C THE PASSERINES

Family	Habitat
Broadbill (Eurylaimidae)	arboreal
Pitta (Pittidae)	ground (in forest)
Larks (Alaudidae)	ground
Swallows (Hirundinidae)	aerial, cliff, bank, building
Cuckooshrikes (Campephagidae)	arboreal
Drongoes (Dicruridae)	arboreal
Orioles (Oriolidae)	arboreal
Crows and **Raven** (Corvidae)	arboreal, ground, cliff
Tits (Paridae)	arboreal
Penduline Tits (Remizidae)	arboreal
Creeper (Salpornithidae)	arboreal
Babblers (Timaliidae)	arboreal, ground
Bulbuls (Pycnonotidae)	arboreal, ground
Thrushes, Robins, Chats (Turdidae)	arboreal, ground
Warblers (Sylviidae)	arboreal, grassland, marsh
Prinias (Sylviidae)	arboreal
Flycatchers (Muscicapidae)	arboreal
Wagtails (Motacillidae)	aquatic
Pipits & Longclaws (Motacillidae)	ground
Shrikes (Laniidae)	arboreal
Starlings (Sturnidae)	arboreal, ground, cliff
Oxpeckers (Buphagidae)	arboreal, large mammal
Sugarbirds (Promeropidae)	arboreal
Sunbirds (Nectariniidae)	arboreal
White-eyes (Zosteropidae)	arboreal
Weavers, Widows, Bishops, Finches, Sparrows, (Ploceidae)	arboreal, grassland, marsh
Whydahs (Viduidae)	arboreal, ground
Waxbills (Estrildidae)	arboreal, ground
Canaries and **Buntings** (Fringillidae)	arboreal, ground, cliff

'Slender-billed arboreal Passerines'

Once you have found your way around *Roberts'* and around the countryside, it remains to find your way around a bird. This is quite simply done by looking at the line drawing of a bird in *Roberts'* on page xlix in the prelims. The names of the parts of a bird speak for themselves; some of the names have synonyms of which the most important are:

maxilla = upper jaw
mandible = lower jaw
forehead = front
belly = abdomen
breast = chest
cheek = malar region, malar stripe, or moustache.

The word 'front' must be remembered because most people tend to think of it as meaning the chest or underside of the bird, but it is never used in this sense; it always means the forehead. The 'culmen' is the ridge along the top of the bill, not the bill itself.

Finally a word about colours and markings of birds. The question of colour is often a difficult one for two main reasons. Firstly the light may affect colour quite drastically. Secondly every person sees colour slightly differently from everyone else (not to mention colourblindness which is a great handicap in bird identification). Even so, most bright colours present few problems. Red, green, yellow, blue, black, white and brown are often straightforward. The problems arise with the use of words like buff, rufous, umber and so on. Not only may people see these colours differently, but they may even interpret them differently. 'Rufous' is a particularly overworked word and covers such colour tones as reddish-brown, rust, russet, sepia,

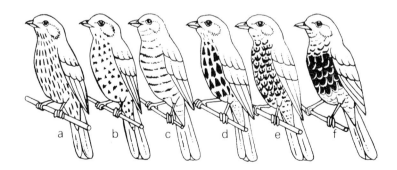

Figure 2   Common markings of birds, often useful in identification:
a = streaked; b = spotted; c = barred; d = blotched (or very heavily spotted); e = mottled; f = scaled (or scalloped).

mahogany, chestnut, auburn, ginger, cinnamon, and even 'red' itself. Only usage will familiarize you with an author's meaning of rufous.

'Buff' on the other hand is usually more on the yellowish side of the spectrum and may include light shades of biscuit, cream, *café au lait*, certain shades of dull pink and yellow, and even paler tones of rufous! Once again, familiarity with a particular book will accustom you to the way in which an author sees and describes his colours.

Markings fall into six main kinds: streaked (or striped), spotted, barred, blotched, mottled, and scaled (or scalloped). These are shown in Figure 2.

Once you've got a good look at your bird, make a sketch and jot down the colours and any other features you have particularly noted – things that have struck you as being obvious and therefore probably useful in identification. These include size, general shape, length and proportions of bill, neck and legs, length and shape of tail, general posture and so on. Once you start using the keys, you will begin to get an idea of the kind of features that are most useful in bird identification. These features are known for this reason as 'key characters'.

Chapter Three

The Waterbirds

From what you learned in Chapter 2, it will be obvious that all the waterbirds, with the exception of the wagtails, are non-passerine and therefore to be looked for only in the first third of *Roberts'*. This is a great help because it automatically saves you having to consider the other two-thirds of the book at all when you are birdwatching along the shoreline of a dam, on the coast or wherever birds feed at or in the water. Waterbirds are mostly fairly large and easy to see, which is why I usually start beginners off with them.

Let us now try to use the key and see how it actually works. Let's assume we are at the waterside of a large inland dam. We see a very large, long-legged wading bird with a long neck and a long, straight, pointed bill. Its coloration is rather dark, mostly slaty grey and deep chestnut. Now let's go to the waterbird key (pages 22–23); start at number 1 where the choice is between inland waters and marine waters. If you are at a dam, you choose inland waters, so run your eye along the line and it will come to rest at '2' at the right hand margin. Go to '2' in the left hand margin and your choice is a threefold one between swimming, wading and flying. Since we are looking at a long-legged wading bird, our choice must go to '6'. At '6' we have a size choice; I said initially that our bird was very large, so it is almost certainly larger than a domestic fowl, which leads us to point 7. The bill of our bird is straight and we are led to point 8. Its plumage is not entirely white, so it must be either one of the larger herons, or a stork.

Find the herons by looking at the list of colour plates on page vii or in the Index in *Roberts'*. The larger herons are on Plate 5, and there in the bottom left hand corner is the dark slaty and chestnut heron we have been looking at. Goliath Heron is the answer to our identification problem. And that's how the key works. Now try some other waterbirds and key them out. The more practice you have with a key, the quicker you'll learn to work with it.

If you look at point 16 on the Waterbird Key, you'll see the word '**Waders**' among those listed. There is an important technical difference between the waders (as a systematic group of related birds) and the wading birds (any other bird that wades in the course of earning its living). The waders are marked in the non-passerine list on page 6 (Table 1). *Roberts'* includes 9 families in this category. A funny thing about the 'waders', used in this sense, is that not all of them wade. Some of them, like the coursers, are desert birds, while many of the plovers are grassland birds which never go near water; yet technically they are waders. In a narrower sense, the term may be applied to the Scolopacidae only, a difficult family to identify; the key to the scolopacid waders on pages 24–27 will help you to key them out.

Chapter Four

Slender-billed Arboreal Passerines

Having dealt with the waterbirds, a large but relatively easy group, let's have a look at the next largest habitat-group of bird families. This is the group that I have termed the arboreal passerines (Table 1). Before we start on them, we can eliminate some families to make the job a little easier. Firstly, let's eliminate the pittas, larks and swallows because they are not arboreal in that they are not found in trees more often than in any other kind of habitat. Then we can eliminate the last four families of passerines, the Ploceidae Viduidae, Estrildidae and Fringillidae, because they have characteristically conical bills (figure 3a – e) and are mostly fairly easy to identify. Exceptions to this rule are the females of most weavers, widows and bishops which give even the experts some identification problems! Finally we can eliminate the crows and the ravens (big, black and unmistakable), the wagtails (mostly waterside birds), and the pipits and longclaws (usually seen on the ground in grassland or savanna habitat).

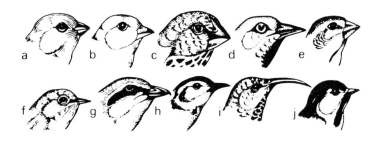

Figure 3   **A selection of different passerine birds to show the two basic bill shapes referred to in the text:**
a—e = conical-billed passerines; f—j = slender-billed passerines.

This narrows down the field to the rest of the passerines, which we can call the 'slender-billed arboreal passerines'. Their bills are only relatively slender in some cases, but the general types can be seen in Figure 3f – j. This still leaves us with 18 families of birds and a lot of species, but the Slender-billed Arboreal Passerine Key (pages 28–29) will help you to sort them out. Practise using the key by choosing an illustration on one of the relevant plates in *Roberts'* and keying it out. You'll first have to check in the text for the size of that species to put you on the right track, because the choice at point 1 in the key is based on size.

If you still find the number of birds daunting, just hold together pages 471–706 of your copy of *Roberts'* and you'll realise that it is only 30% of the book. The waterbirds were 26% of the book, so the passerines in the slender-billed arboreal category are not many more to deal with.

Chapter Five

The Near-Passerines

Technically the near-passerines are of course just non-passerines (see Table 1) since they are not passerines, but they form a sort of 'shadow-group' between birds that are obviously non-passerines and birds that are definitely passerines. Don't worry about the definitions of these major groupings; you will get the feel of them with experience. The distinctions between them are in any case rather deeply scientific.

From the list of near-passerines on page 7 you can see that by far the greater majority are arboreal. This is another reason why a beginner might think he was looking at a passerine. If you are looking at a smallish bird in a tree, that won't key out in the slender-billed arboreal passerine key, or that is not one of the conical-billed passerines (weavers, canaries, etc.), try keying it out in the near-passerine key on pages 30–31.

Near-passerines are generally an easy group of birds to identify because of a characteristic body and face shape, like the owls. Only a few, such as the nightjars, honeyguides, woodpeckers and swifts, present real identification problems, since they tend to be rather dull in colour and therefore somewhat similar in appearance. (At this point, try to learn the difference between swifts and swallows by studying carefully Plates 39 and 46 in *Roberts'*.) Only the cuckoos can be considered a difficult group on habits alone; they tend to sit among dense leaves near the tops of trees as they sing, but fortunately their songs are so easy to identify that the problem almost solves itself before it becomes too discouraging.

Woodpeckers in southern Africa are mostly very similar, but a careful look will usually enable one to identify them easily with the aid of the keys on pages 32–34.

For our purposes the near-passerines comprise a mere 15% of *Roberts'*, or half as many as the slender-billed arboreal passerines. Again, practise by using the key with representatives in the field and in *Roberts'*.

Chapter Six

Five Tricky Groups

You have now reached a point at which you are fairly familiar with the names of all the major groups of birds to be found in southern Africa. You should be able to use a key without any trouble and you will be getting the idea of key characters. These are the characters one uses in the field to separate birds into groups and eventually to identify them to species. So you can now be a bit more adventurous!

I have chosen five groups of birds that present problems of field identification to everyone, not only to the beginner. These are the swallows, the francolins, the larks, the cisticolas and the eclipse male and female widows, in ascending order of difficulty. All but the cisticolas and widows are complete families on their own. The cisticolas are all warblers in the genus *Cisticola*, and they are all small, brown quick-moving and elusive. The widows belong to the genus *Euplectes*, one of the weaver genera; I have also included the very similar *Quelea* females.

The Swallows

Let's start with a relatively easy group, the swallows. All are distinctively marked but, because they are often (if not usually) in flight, they may be hard to see properly for long periods at a time. So the key to the swallows (pages 35–36) is designed to help you to pick up features quickly, so that you can identify any swallow or martin with certainty. This key omits the East African Sawwing Swallow, which is rare in southern Africa, but is in any case very easy to identify by its black body and white underwing pattern. Try to see swallows in good light (preferably bright sunlight) before attempting to key them out.

The Francolins

Francolins present totally different identification problems from the swallows. They are ground birds that skulk away into dense vegetation, or take wing and disappear when disturbed. They are therefore often hard to see well enough to be able to note down their potentially diagnostic features. But they have their own advantages. Firstly they are a good size. Secondly they are easily divisible into a red-billed group and a dark-billed group. Thirdly they *do* have distinctive markings, although it's usually hard to see belly markings because the birds tend to crouch as they walk or run away from one. Finally francolins all have distinctive calls by which their field identification can be confirmed.

If you are visiting a game reserve where francolins are often tame at the roadsides, practise using the key (pages 37–38) on them. You'll be surprised at how easy it is when approached in this simple series of steps.

The Larks

Ask any birder, amateur or professional, experienced or inexperienced, which group of birds presents the greatest identification problems. The answer is more than likely to be 'the larks'. I am glad to say that this is something of an exaggeration. I am not saying that the larks are easy, but they are not the impossibly difficult group they are made out to be. Given a good look at the bird and a chance to jot down the features in your notebook, you ought to be able to identify almost any lark. Let me stress again the need to *write down* all the relevant information on the spot; don't try to remember it, because you simply won't be able to recall all the necessary details later on. Use the lark key (pages 48–50) and then look up Plates 44 and 45 in *Roberts'* to confirm your identification.

Larks, like cisticolas and most francolins, also have distinctive songs and some have fairly spectacular display flights to accompany them. So you can usually confirm your provisional sight-identification by looking up the song in *Roberts* .

If your lark won't key out, or else does not seem 'quite right', it is possible that you may have a pipit instead. The pipits are a notoriously difficult group of birds. Even the experts find it hard to separate many species in the field, expecially the open-grassland species. Pipits all have white or buff outer tail feathers. So do some larks, like the Melodious Lark, Monotonous Lark and Redcapped Lark, but the larks usually have feathers that help to separate them from the pipits (such things as rufous-edged wing feathers, rufous cap, more spotty on the back, and so on). Look carefully at the larks in the plates and compare them with the pipits on Plate 59.

The Cisticolas

Cisticolas (pronounced Sis-*tik*-ola) are surely among the hardest of the southern African 'LBJs' (Little Brown Jobs) to identify. A key based on plumage features is almost useless because these little warblers look so much alike though the key to the cisticolas on pages 51–53 will certainly make your job a lot easier. Just look at Plates 55 and 56 in *Roberts'* to see what I mean. So I've adopted a quite different approach to the cisticolas. Look at Table 2 on page 17. The left hand column lists the 17 southern African species of cisticola. Along the top of the table are two sets of distribution patterns separated by a heavy vertical line: the one set gives the distribution by altitude above sea level, and is very approximate, because the further into the tropics one goes, the higher will the corresponding altitude be, and of course the further south one moves, the closer to sea level will the species occur. The other set of distributions is by habitats. The way to sort out the most likely cisticolas to occur in a given area is to use the altitudinal and habitat information together.

For example, if you are in grassveld in the Orange Free State, put one finger on the 'Highveld over 1200 metres' column and another finger on the 'Open short grassveld' column. Now run the two fingers down the columns; wherever a + occurs in *both* columns opposite the same species (that is to say on the same horizontal line), that species is likely to occur in that region and habitat. The result of this particular exercise is the following list of species: Fantailed Cisticola, Desert Cisticola, Cloud Cisticola and Ayres' Cisticola. Adjacent marshland may harbour Levaillant's Cisticola, and nearby scrub may have the Neddicky, but if you are strict in your application of the habitat rule, your list can never be very long.

Now let's try another area and a different habitat — how about 'Lowveld under 500 metres' and 'Marshy or streamside vegetation'. The result? Fantailed Cisticola, Palecrowned Cisticola, Wailing Cisticola, Redfaced Cisticola, Blackbacked Cisticola, Chirping Cisticola, Levaillant's Cisticola and Croaking Cisticola. This may seem a rather long list of eight potential LBJs, but it is highly unlikely that in any given marsh in the low-lying areas of southern Africa you will find more that three or four of these species side by side. Also, all of these eight species of cisticola can be quite easily identified by song, and several of them are fairly distinctive in appearance too.

A breakdown of Table 2 will show that the greatest possible number of cisticola species in any habitat is eight (the example just cited). One other habitat has a possible seven species (Tall grassveld at Middleveld altitudes), three have a possible five species, five have a possible four species and all the rest have three species or fewer.

If you can get a good look at a cisticola (not always easy), you will see that it has the crown either streaked black on tawny, or plain rusty red. The back may be marked with bold or light streaks, or not at all. Sometimes the markings may be so bold as to make the back almost solid black. Check the illustrations in *Roberts'* for these combinations of characters. However, one of the safest ways of identifying cisticolas is by song. The descriptions of cisticola songs in *Roberts'* are very clear, aided as they are by sonagrams, and you can use combinations of song, habitat, distribution and appearance to get the best result. In fact the principles of identification you will need to use here are the same as for any other group of birds, but you'll just have to work a bit harder to get the information you need.

Table 2: Cisticola Distribution and Habitats

(*Roberts'* pages 578–595; Plates 55–56)

Species	'Lowveld' (under 500 m a.s.l.)	'Middleveld' (500 – 1200 m a.s.l.)	'Highveld' (over 1200 m a.s.l.)	Bushveld	Savanna	Shrub or scrub	Rank vegetation (weeds, etc.)	Tall grassveld	Marshy or streamside vegetation	Open rank grassveld	Open short grassveld	Cultivated fields	Gardens	Mountains and rocks	Desert and semidesert
Fantailed	+	+	+					+	+	+	+	+			+
Desert			+							+	+				+
Cloud		+	+								+				
Ayres'		+	+					+			+			+	
Palecrowned	+	+							+	+					
Shortwinged	+			+	+										
Neddicky	+	+	+	+	+	+							+	+	
Greybacked	+	+				+								+	
Wailing	+	+	+				+	+	+					+	
Tinkling	+	+	+		+										+
Rattling	+	+		+	+	+		+							
Singing	+			+	+	+	+	+							
Redfaced	+			+			+		+						
Blackbacked	+						+		+						
Chirping	+	+							+						
Levaillant's	+	+	+				+	+	+						
Croaking	+	+		+				+	+	+					
Lazy	+	+	+	+			+	+	+					+	

The female and eclipse *Euplectes* and *Quelea* species

Take a look at the bishops and widows on *Roberts*' Plate 67 and the queleas at the top of Plate 68. It is immediately obvious that the males in breeding (or nuptial) plumage are startlingly different from each other in their black, red and yellow colours, while the females and eclipse males are as startlingly similar to each other in their streaky or mottled browns and buffs. In the summer, when the males are in breeding condition, the streaky brown birds in a flock can usually safely be assumed to belong to the same species as the males in that flock, but in winter not only do the males lose their bright coloration and assume what is known as eclipse plumage which resembles very much that of the females, but the different species often form mixed flocks.

Although the females and eclipse males look very much alike, they are by no means impossible to identify, but, as with the larks, one must use critical observation and rather fine detail. The key on pages 54–55 will show you what criteria to use in sorting out this confusing group of look-alikes. I would recommend also that you read the article by D. N. Johnson and R. F. Horner entitled 'Identifying widows, bishops and queleas in female plumage' (1987, *Bokmakierie* 38: 13–17) as it gives many more tips about identifying these birds.

Chapter Seven

The Raptors, or Diurnal Birds of Prey

Now that you have dealt with the waterbirds and the francolins, very few non-passerines are left to be identified. The largest remaining group is the raptors, or birds of prey (excluding the owls, or nocturnal birds of prey). These cover only 3% of all the birds in *Roberts*', but include some of the most conspicuous birds that you will encounter in your travels. But at the same time as being conspicuous, the raptors are among the most difficult birds to identify in the field. Plumages may differ very little between certain different species, yet they may differ markedly between age-groups of a single species, and sometimes the sexes may differ too. Add to this individual plumage variations, colour phases (e.g. dark and light) and geographical plumage differences and you get a very complicated picture indeed. Even so, few of them are impossible to sort out.

The Raptor Key (pages 39–47) is the most extensive single key in this booklet, because it attempts to key out every major plumage variation among raptors in southern Africa, whether based on species, sex, phase, age or distribution, sometimes even allowing for the more important individual variations. Since many raptors are first seen from below, I have selected belly colour as a starting point for all raptors in the key. The choice is initially between paler colours and darker colours. If in doubt, try one choice and then the other to see what answers you get. If you are lucky, you may get to the same answer in both cases!

Several raptors (especially the goshawks) have a plain grey or brown chest and a barred belly. The bars may be white and grey, or grey and brown, or white and brown. In the key you will sometimes be required to decide on whether the chest and belly colours are the same or contrasting. To make your decision simple, the rule for purposes of this key is: *if the belly is barred with equal widths of two colours, the lighter colour is the background colour.* This is explained by means of the diagrams in Figure 4. For example, look at the adults of both Dark and Pale Chanting Goshawks on Plate 16 in *Roberts'*. The chest in each case is plain grey, the belly barred grey and white. Since the

Figure 4   **Three goshawks showing different types of barring:**
a = barred grey on white (bars of equal widths);
b = barred white on grey (white bars narrower than grey bars);
c = barred grey on white (grey bars narrower than white bars).

barring is equal widths of grey and white, the ground colour is white, not grey. So in both these cases the belly colour contrasts with the chest colour (as will be the case in nearly all species with a plain chest and a barred belly).

The length of the raptor key, and therefore the time it may take you to key out your bird makes it imperative to write down as many features as possible while you are actually looking at the bird in the field, before you start to use the key, just in case the bird flies away. Always work on the principle that you are not likely to get a second look at any bird. Another possibility, if there are two people together in the field, is to have one person reading out from the key while the other looks at the bird, but this is not as safe a method as writing it all down first.

Chapter Eight

A Word of Encouragement

The keys you have worked through so far cover about 82% of the birds in *Roberts'* and I have mentioned some other groups in passing. Those that remain are mostly easy to identify (guineafowl, buttonquails, sandgrouse, coursers, pratincoles, doves, chats and the conical-billed seedeating passerines) and you shouldn't have much trouble with them. Nobody can identify every single bird species with complete certainty. The pipits continue to baffle the experts, and the little brown birds still tend to disappear before you have had a really good look at them. So don't worry if some birds elude you. The chase is often more fun than the kill, so enjoy your spotting and cut your losses on those you can't pin down. There is a saying among museum ornithologists: 'What's hit is history, what's missed is mystery'. A little mystery never did anyone any harm.

To end off with I have added a few keys to the commoner species-groups and species-pairs. These are groups of closely similar species that sometimes present problems of identification because their separation often rests on one or two finer details which need to be looked for specially.

No key is completely foolproof, nor is any key the last word on the subject of identifying a particular group. Any group may be keyed out by one of several different keys, each adopting a different approach. One way in which to improve your powers of observation and your ability to use a key is to make one yourself. Select a fairly straightforward group of birds to start with – something like the double-collared sunbirds on Plate 63 in *Roberts'*. Then you can become more ambitious and make your own key to the conical-billed passerines (Plates 65–72), either by families or as a whole. You will be amazed at how your eye for detail is sharpened by having to pick out key characters.

One final word. Birds are living things that undergo changes in their appearance according to age, season or sex. Keep this in mind all the time. Be prepared for any eventuality. Read every word of the text in *Roberts' birds of southern Africa* for each species you are trying to sort out. Check the distribution map to see whether it ought to occur in your area. It is always safest to assume that the bird you are looking at is not rare and that it is in its right habitat and distribution: so avoid the temptation to claim an unusual find, especially if you are a beginner.

Happy birdwatching!

(*Roberts'* pages 2–297)

1 Inland waters (lakes, dams, pans, rivers, streams) 2
   Marine waters 12

2 Swimming (or diving from surface of water) 3
   Wading on or near shoreline (mostly long-legged) 6
   Flying, or diving from air (mostly long-winged) . . . .
   . . . . **Terns (Pl. 32), Skimmer (Pl. 31)**

3 As large as turkey, or larger 4
   Smaller than turkey 5

4 Mainly white . . . . **Pelicans (Pl. 4)**
   Mainly black . . . . **Spurwinged Goose (Pl. 8)**

5 Bill pointed (not hooked at tip) . . . .
      Grebes (Pl. 6), Darter (Pl. 4), Coot, Moorhen, Finfoot (Pl. 22)
   Bill blunt or hooked at tip . . . .
   . . . . **Cormorants (Pl. 4), Ducks (Pls 8–9), Gulls (Pl. 31)**

6 Larger than domestic fowl 7
   Smaller than domestic fowl (or about the same size) 10

7 Bill straight 8
   Bill curved 9

8 Plumage entirely white . . . .
   . . . . **Great White Egret (Pl. 5), Spoonbill (Pl. 7)**
   Plumage not entirely white . . **Larger Herons (Pl. 5), Storks (Pl. 7)**

9 Bill short; legs and neck very long . . . . **Flamingoes (Pl. 7)**
   Bill long . . . . **Ibises (Pl. 7), Curlew (Pl. 25)**

10 Plumage boldly black and white ....
.... **Blacksmith Plover (Pl. 26), Avocet, Stilt (Pl. 25),**
Pied Wagtail (Pl. 59)
Plumage not boldly black and white 11

11 Uniform dark brown .... **Hamerkop (Pl. 7)**
Not uniform dark brown 16

12 Swimming (or diving from surface of water) 13
Wading on or near shoreline (mostly long-legged) 14
Flying, or diving from air (mostly long-winged) ....
..... **Albatrosses (Pl. 2), Petrels (Pl. 3), Gannets, Frigatebirds,**
Tropicbirds (Pl. 4), Terns (Pl. 32)

13 Open sea .... **Penguins (Pl. 4), Albatrosses (Pl. 2), Petrels (Pl. 3),**
Phalaropes (Pl. 29)
Inshore (within about 4 km of land) .. **Blacknecked Grebe (Pl. 6),**
Pelicans, Gannets, Cormorants (Pl. 4), Gulls (Pl. 31)

14 Mainly black (belly may be white); bill red ....
.... **Oystercatchers (Pl. 25)**
Not mainly black 15

15 Larger than domestic fowl ... **Flamingoes (Pl. 7), Curlew (Pl. 25)**
As large as, or smaller than domestic fowl .... **Turnstone,**
Plovers (Pl. 26), 'Waders' (Pl. 29),* Avocet,
Crab Plover (Pl. 25), Gulls (Pl. 31), Wagtails (Pl. 59)

16 Bill markedly longer than head ..... **Some small Herons (Pl. 6),**
African Rail (Pl. 22), Painted Snipe, Snipe (Pl. 25),
some 'Waders' (= Scolopacidae)* (Pls 25 & 29)
Bill as long as, or shorter than, head ....
Some small Herons (Pl. 6), Rails, Crakes (Pl. 22),
Jacanas (Pl. 23), Plovers (Pl, 26), many smaller 'Waders'
(= Scolopacidae)* (Pl. 29), Water Dikkop (Pl. 25),
Wagtails (Pl. 59)

* See key to scolopacid waders on p. 24.

KEY TO THE SCOLOPACID WADERS
(SANDPIPERS AND THEIR ALLIES)

(*Roberts'* pages 235–258; Plates 25, 28–29, & 74)

Every summer the southern African avifauna is augmented by a large influx of waders from the northern hemisphere. These are birds of the family Scolopacidae, all of which breed in the northern hemisphere, mostly in the tundra, and migrate to the southern hemisphere in order to spend their nonbreeding season in a mild climate. These waders tend to be rather dull-coloured for the most part, and therefore rather similar in appearance — especially the smaller ones with relatively short bills. They are found largely on the shores of lakes and the sea, wading at the water's edge and probing the mud for food. Because they constitute such a large group and present some special identification problems, I have provided the 'wader key' to try to sort these out for you. It is important to combine observations of the bird on the ground and in flight whenever possible, voice is a useful additional identification aid.

1   Size large (over 40 cm); bill much longer than head, markedly decurved 2
     Size medium to small (under 40 cm); if large, then bill straight or slightly upcurved 3

2   Eyebrow distinctly pale; one pale and two dark lines on crown; rump white . . . . **Whimbrel (Pl. 25)**
     Head indistinctly streaky; rump and lower back conspicuously white in flight ˙. . . . . **Curlew (Pl. 25)**

3   Bill noticeably longer than head, straight or slightly upcurved . .   4
     Bill as long as, or shorter than, head, or markedly decurved . .   11

4   Bill pink or orange at base, dark towards tip; dorsal plumage looks plain at distance; legs long, slender . . . . . . ·. 5
     Bill uniformly dark throughout, or dull yellowish 7

5   Size larger (around 40 cm); legs dark 6
     Size smaller (around 25 cm); legs orange-red 22

6 Tail white with broad black tip; broad white wingbar in flight . . .
. . . . **Blacktailed Godwit (Pl. 25)**
Tail white, narrowly barred black; no wingbar in flight . . . .
. . . . **Bartailed Godwit (Pl. 25)**

7 Head and back heavily streaked buff on dark brown; legs rather
short; skulk in dense marshy habitat 8
Head and back look plain greyish or with faint pattern only; legs
long, greenish or yellowish; wade in open water 9

8 Belly buff with black chevrons; silent when flushed . . . .
. . . . **Great Snipe (Pl. 25)**
Belly white with black bars; calls when flushed . . . .
. . . . **Ethiopian Snipe (Pl. 25)**

9 Bill markedly upcurved, orange at base; legs yellow, rather short;
dark patch on bend of wing; trailing edge of wing white;
rump and tail grey . . . . **Terek Sandpiper (Pls 28, 29)**
Bill all dark, straight or only slightly upcurved; legs greenish, long;
no white in wing; rump and lower back look white in flight . .
. . . . 10

10 Bill thin and tapering to tip; build slender (length about 23 cm);
callnote single *chuk* . . . . **Marsh Sandpiper (Pls 28, 29)**
Bill relatively stout, not tapering to tip, sometimes slightly upcurved;
build robust (length about 32 cm); call loud 3-note whistled
tew-tew-tew . . . . **Greenshank (Pls 28, 29)**

11 Bill straight or nearly straight 14
Bill decurved at tip 12

12 Bold white eyebrow forks behind eye; dark patch on bend of wing;
sides of chest streaky; legs rather short . . . .
. . . . **Broadbilled Sandpiper (Pls 28, 29)**
Eyebrow pale, but not bold, nor forking behind eye; wings and
back uniform grey; sides of chest washed grey 13

13 Rump plain white in flight . . . . . . **Curlew Sandpiper (Pls 28, 29)**
    Rump white with black centre . . . . **Dunlin (Pls 28, 29)**

14 Back boldly scaled buff; no pale eyebrow; legs usually orange or
    yellowish 23
    Back plain or variously patterned; eyebrow pale; bill always
    short and straight 15

15 Back looks plain dark brown; chest brownish; white of belly ex-
    tends up around bend of wing; bobs frequently . . . .
    . . . . **Common Sandpiper (Pls 28, 29)**
    Back spotted or mottled, never plain 16

16 Chest plain white like belly 17
    Chest brownish or greyish, contrasting with white belly 19

17 Overall appearance very pale; back pale grey, faintly mottled; bill
    and legs blackish; dark patch at bend of wing; broad white
    wingbar in flight; almost entirely marine shores; runs fast . . .
    . . . . **Sanderling (Pls 28, 29)**
    Back grey or brownish; size very small; narrow white wingbar in
    flight; often on inland waters 18

18 Back heavily mottled, brownish; common, gregarious . . . .
    . . . . **Little Stint (Pls 28, 29)**
    Back lightly mottled, greyish; rare vagrant; flocks small . . . .
    . . . . **Rednecked Stint (Pls 28, 29)**

19 Back brown or buffy brown; no white wingbar in flight 20
    Back grey, mottled or scaled; bold white wingbar in flight; rump
    pale, mottled; bill and legs black; build chunky . . . .
    . . . . **Knot (Pls 28, 29)**

20 Back buffy brown, mottled; brownish, streaked chest ends abruptly
    at white belly; rump pale with dark centre line . . . .
    . . . . **Pectoral Sandpiper (Pls 28, 29)**
    Back dark brown, spotted or speckled with white; rump white;
    chest coloration merges into white of belly 21

21 Back spotted with white; chest greyish or brownish, merging into
     white of belly; legs yellowish green; underwing pale . . . .
     . . . . **Wood Sandpiper (Pls 28, 29)**
     Back finely speckled with white; chest brownish, contrasting more
     markedly with white of belly; legs dark; underwing looks
     black . . . . **Green Sandpiper (Pls 28, 29)**

22 Back faintly streaked; trailing edge of wing broadly white; call
     loud *tewk-tewk-tewk* . . . . **Redshank (Pls 28, 29)**
     Back almost plain; dark wings contrast with grey back; trailing
     edge of wing darkly mottled (not white); call liquid *tew-it* . . .
     . . . . **Spotted Redshank (Pl. 74)**

23 Pale ring around base of bill; narrow white wingbar in flight; belly
     white; rump white with dark centre line; legs usually dull
     orange (reddish to yellowish); usually silent . . . .
     . . . . **Ruff (Pls 28, 29)**
     Eyering white; underparts uniform deep buff; no wingbar in flight;
     rump mottled like back; underwing pure white; legs dull
     yellow . . . . **Buffbreasted Sandpiper (Pl. 74)**

(*Roberts'* pages 471–706)

1 Very small birds (less than 13 cm; smaller than sparrow) 2
  Small to medium-sized birds (longer than 13 cm) 4

2 Bill long and decurved . . . . **Sunbirds (Pls 63–64)**
  Bill relatively short 3

3 Body mostly yellow or green; eye-ring white . . **White-eyes (Pl. 64)**
  Body not mostly yellow or green . . . . **Warblers (Pls 52–56),**
      some Flycatchers (Pls 57–58), Penduline Tits (Pl. 53)

4 Body all black, sometimes with blue or green gloss . . . .
      . . . . **Male Black Cuckooshrike, Drongoes (Pl. 47),**
          Black Flycatcher (Pl. 57), some Starlings (Pl. 62)
  Body not all black 5

5 Body bright yellow above and below; wings black . . . .
      . . . . **Orioles (Pl. 47)**
  Body not bright yellow above 6

6 Bill long and decurved . . . . **Sugarbirds (Pl. 62),**
      Sunbirds (Pls 63–64), Spotted Creeper (Pl. 48)
  Bill relatively short 7

7 Tail orange with black centre . . . .
      . . . . **Some Robins and Chats (Pls 50–51)**
  Tail not as above 8

8 Heavily streaked or spotted below . . . . **Broadbill (Pl. 36),**
      some Thrushes (Pl. 49), Whitebrowed Robin (Pl. 51)
      female Plumcoloured Starling (Pl. 62), Striped Pipit (Pl. 59)
  Not heavily streaked or spotted below 9

9 Tail about twice body-length . . . .
      . . . . . **Longtailed Shrike (Pl. 60), Paradise Flycatcher (Pl. 58)**
  Tail not very long 10

10 Bright yellow below . . . . **Stripecheeked Bulbul,**
   Yellowbellied Bulbul (Pl. 48), Starred Robin (Pl. 51),
   some Shrikes (Pl. 61)
   Not bright yellow below 11

11 Plain dove-grey all over . . . . **Grey Cuckooshrike (Pl. 47)**
   Not as above, or dull olive-grey all over 12

12 Bill bright yellow, orange or red . . . . . . **Some Thrushes (Pl. 49),**
   some Helmetshrikes (Pl. 60), Oxpeckers,
   European Starling, Indian Myna (Pl. 62)
   Bill black or dull coloured 13

13 Belly pure white . . . . **Whitebreasted Cuckooshrike (Pl. 47),**
   Pied & Barecheeked Babblers (Pl. 49),
   some Flycatchers (Pls 57–58), some Shrikes (Pl. 60),
   male Plumcoloured Starling (Pl. 62)
   Belly coloured or off-white 14

14 Belly black . . . . **Black Tit (Pl. 48), Arnot's Chat (Pl. 50)**
   Belly not black 15

15 Head black, with or without eye-stripe . . . . **Tits (Pl. 48),**
   Redeyed, Cape and Blackeyed Bulbuls (Pl. 48),
   some Flycatchers (Pls 57–58), some Shrikes (Pls 60–61)
   Head not black 16

16 Finely barred below . . . . . . **Female Black Cuckooshrike (Pl. 47),**
   female Redbacked Shrike (Pl. 60), some immature Shrikes
   Not finely barred below 17

17 Wings bright chestnut . . . . **Tchagras (Pl. 61)**
   Wings not chestnut 18

18 Gregarious; noisy . . . . **Starlings (Pl. 62), Babblers (Pl. 49),**
   Bulbuls (Pl. 48), Helmetshrikes (Pl. 60)
   Solitary; quiet . . . . **Thrushes (Pl. 49), Robins (Pl. 51),**
   Flycatchers (Pls 57–58), Shrikes (Pls 60–61)

(*Roberts'* pages 314—424)

1 Bill large, decurved, sometimes topped by a casque . . . .
. . . . **Hornbills (Pl. 41)**
  Bill relatively smaller, but straight if large   2

2 Bill large, long, straight and pointed . . . . **Kingfishers (Pl. 40)**
  Bill relatively short, but slender if long   3

3 Bill long, slender and curved . . . .
. . . . **Hoopoes (Pl. 34), most Bee-eaters (Pl. 40)**
  Bill relatively shorter and/or stouter   4

4 Head markedly crested   5
  Head not markedly crested   6

5 Birds small; tail long and pointed . . . . **Mousebirds (Pl. 42)**
  Birds medium to large; tail rounded . . . . **Louries (Pl. 34),**
        Crested Barbet (Pl. 42), some larger Cuckoos (Pl. 35)

6 Bill very short, stout and hooked . . **Parrots (Pl. 34), Owls (Pl. 37)**
  Bill not as above   7

7 Bill straight and chisel-like . . . . **Woodpeckers,* Wryneck (Pl. 43)**
  Bill strongly or slightly curved on top, or very small   8

8 Outer tail feathers strikingly white in flight . . . . . . **Honeyguides,**
        Klaas's Cuckoo (Pl. 36), some Nightjars (Pl. 38)
  Outer tail feathers not white, or with white spots only   9

9 Birds camouflaged, small-billed, long-winged . . . **Nightjars (Pl. 38)**
  Birds not as above 10

* See woodpecker keys on pages 32–34.

10 Back green; belly red, orange or dull yellow 16
   Not as above 11

11 Wings chestnut; tail heavy and dark; bill stout . . . .
   . . . . **Coucals (Pl. 35)**
   Not as above 12

12 Bill bright yellow; body drab greenish . . . . **Green Coucal (Pl. 35)**
   Bill black, or yellow at base only 13

13 Medium-sized; wings blue; perch conspicuously . . . .
   . . . . **Rollers (Pl. 41)**
   Smaller; not as above 14

14 Birds small to very small; bill heavy or stubby; head may be
   striped black and white . . . . **Barbets (Pl. 42)**
   Birds not as above 15

15 Aerial, long-winged; mainly black or grey . . . . **Swifts (Pl. 39)**
   Arboreal; medium to small; fast fliers; shy . . . .
   . . . . **Cuckoos (Pls 35–36)**

16 Back dark green; belly bright red . . . . . . **Narina Trogon (Pl. 34)**
   Back light green; belly orange or dull yellow; collar black . . . . . .
   . . . . **Little Bee-eater (Pl. 40)**

(*Roberts'* pages 416–423; Plate 43)

All southern African woodpeckers, except the Ground and Olive Woodpeckers are clearly marked on the belly. Streaked bellies are found in Goldentailed, Knysna and Cardinal Woodpeckers. Spotted bellies are found in Bennett's, Specklethroated and Little Spotted Woodpeckers. Only the Bearded Woodpecker has a barred belly. However, since one may not always see the belly, or may need confirmation from other features, the following key is provided.

KEY TO THE WOODPECKERS (MALES)

1 Crown all grey; belly pink to red; rump red; back and wings olive, wings spotted pale yellow . . . .
. . . . **Ground Woodpecker** (♂ and ♀ similar)
Hindcrown red 2

2 Back plain olive green; breast golden; face and forehead grey; rump red . . . . **Olive Woodpecker**
Back barred or spotted; breast spotted, blotched, streaked or barred 3

3 Whole crown red; face white 4
Forecrown not red; or red mottled with black 5

4 Throat plain white . . . . **Bennett's Woodpecker**
Throat finely speckled black on white . . . .
. . . . **Specklethroated Woodpecker**

5 Forecrown plain brown; malar stripe black; breast streaked . . . . .
. . . . **Cardinal Woodpecker**
Forecrown mottled or spotted; malar stripe red or absent; breast streaked, spotted, barred or blotched 6

6 Forecrown mottled red and black; breast spotted, streaked or blotched; malar stripe red or absent 7

Forecrown spotted white on black; breast barred; malar stripe black . . . . **Bearded Woodpecker**

7 No malar stripe; back and breast spotted . . . .
. . . . **Little Spotted Woodpecker**

Malar stripe red 8

8 Back barred; breast streaked or spotted . . . .
. . . . **Goldentailed Woodpecker**

Back spotted; breast blotched . . . . **Knysna Woodpecker**

KEY TO THE WOODPECKERS (FEMALES)

1 Crown all grey; back olive; rump red 2

Forecrown spotted or brown; hindcrown black or red 3

2 Wings spotted pale yellow; breast and belly pinkish red . . . .
. . . . **Ground Woodpecker**

Wings plain olive; breast golden; belly greyish . . . .
. . . . **Olive Woodpecker**

3 Hindcrown red 5

Hindcrown black 4

4 Forecrown brown; breast streaked . . . . . . **Cardinal Woodpecker**

Forecrown spotted white on black; breast barred . . . .
. . . . **Bearded Woodpecker**

5 Back barred 7

Back spotted 6

6 Malar stripe black, speckled white; breast heavily blotched with black . . . . **Knysna Woodpecker**

No malar stripe; breast spotted with black . . . .
. . . . **Little Spotted Woodpecker**

7 Throat and stripe under eye brown; malar region white . . . .
. . . . **Bennett's Woodpecker**

Throat white, finely speckled or streaked with black 8

8 Throat and breast finely speckled . . **Specklethroated Woodpecker**

Throat and breast boldly streaked or spotted . . . .
. . . . **Goldentailed Woodpecker**

(*Roberts'* pages 452–470;   Plate 46)

1   Very small; whole body jet black . . . . . .   **Black Sawwing Swallow**
    Body brown or blueblack above;   below white, rufous, brown or
       blueblack   2

2   Body all blueblack; outer tail feathers very long . . .   **Blue Swallow**
    Body not all blueblack   3

3   Body brown above   4
    Body mostly blueblack above   6

4   Belly pale brown; tail square with white windows . .   **Rock Martin**
    Belly white; no white windows in tail   5

5   Throat white; collar brown   16
    Throat brown; tail slightly forked . . . .   **Brownthroated Martin**

6   Streaked below with black on white   7
    Not streaked below; belly plain white or rufous   8

7   Ventral streaks heavy; ear coverts red like crown . . . .
       . . . .   **Lesser Striped Swallow**
    Ventral streaks light; ear coverts white like throat . . . .
       . . . .   **Greater Striped Swallow**

8   Belly rufous or rufous-buff   9
    Belly white or buffy   10

9   Throat white . . . .   **Mosque Swallow**
    Throat rufous like belly   18

10 Crown red . . . .   **Wiretailed Swallow**
    Crown not red   11

11   Throat dark rufous or buffy (not pure white) 12
     Throat white like belly   13

12   Tail forked, with white windows 17
     Tail square, without white windows . . . . **Cliff Swallow**

13   Rump blueblack like back   14
     Rump grey or white   15

14   White windows in tail; front chestnut; collar complete . . . .
     . . . . **Whitethroated Swallow**
     No white windows in tail; front blueblack like crown; collar
     absent or incomplete . . . . **Pearlbreasted Swallow**

15   Rump and crown grey . . . . **Greyrumped Swallow**
     Rump white; crown blueblack like back . . . . **House Martin**

16   Tail square; eyebrow white   . . . . **Banded Martin**
     Tail slightly forked; no white eyebrow . . . . **Sand Martin**

17   Rufous (or buff) on throat only;   collar broad;   tail deeply forked
     . . . . **European Swallow**
     Rufous on throat and upper breast; collar narrow, broken; tail less
     deeply forked   . . . . **Angola Swallow**

18   Hindneck rufous; no white in tail; underparts paler . . . .
     . . . . **Redrumped Swallow**
     Hindneck blueblack like crown and back;   white windows in tail;
     underparts deeper . . . . **Redbreasted Swallow**

KEY TO THE FRANCOLINS

(*Roberts'* pages 169–178; Plates 20–21)

1 Bill partly or wholly red 2
   Bill without trace of red 6

2 Throat and eye-patch red 3
   Throat not red 4

3 Streaked white on black below; legs red . . . **Rednecked Francolin**
   Streaked black on brown below; legs black . . . .
   . . . . **Swainson's Francolin**

4 Eye-patch yellow; finely barred below . . . . . **Redbilled Francolin**
   No eye-patch; streaked or scaled below 5

5 Belly boldly streaked with white . . . . **Cape Francolin**
   Belly with black and white scaly markings . . . . . . **Natal Francolin**

6 Whole body appears uniform dark brown . . . .
   . . . . **Female Hartlaub's Francolin**
   Variously and distinctively marked above and below 7

7 Boldly streaked brown on white below . . . .
   . . . . **Male Hartlaub's Francolin**
   Not as above 8

8 Head plain yellowish with reddish crown; chest and belly boldly
       barred . . . . **Male Coqui Francolin**
   Head variously marked 9

9 Throat grey . . . . **Greywing Francolin**
   Throat white 10

10 White of throat bordered by black necklace 12
    No black necklace around throat 11

11 Reddish streaks on flanks . . . . 'Kirk's' Francolin (*rovuma*)
    No streaks on flanks; belly finely barred . . . .
    . . . . Crested Francolin (*sephaena*)

12 Chest plain reddish cinnamon; belly boldly barred black on white
    . . . . Female Coqui Francolin
    Chest variously streaked 13

13 Centre of belly barred black and white; chest and flanks streaked
    bright chestnut . . . . Shelley's Francolin
    Belly not barred 14

14 Black-and-white necklace broad, extending to upper chest . . . . . .
    . . . . Redwing Francolin
    Necklace narrow, confined to throat . . . . Orange River Francolin

(*Roberts'* pages 106–187; Plates 10–19)

1 Belly generally pale (white, buff, pale grey) 2
  Belly generally dark (rufous, dark grey, brown, black) 33

2 Belly unmarked (plain) 3
  Belly barred, streaked, spotted or mottled 11

3 Belly grey 81
  Belly white 4
  Belly buff 70

4 Chest white like belly 5
  Chest darker than belly 9

5 Size very small; wings and tail spotted with white; rump white;
    back grey in male, dark brown in female . . . .
    . . . . **Pygmy Falcon (Pl. 19)**
  Size medium to large 6

6 Back plain dove grey 7
  Back dark slate grey, brown, black or mottled 8

7 'Shoulders' black; hovers often . . . **Blackshouldered Kite (Pl. 15)**
  No black on 'shoulders'; sails over ground . . . .
    . . . . **Male Pallid Harrier (Pl. 15)**

8 Size very large; legs feathered to toes . . . .
    . . . . **Immature Martial Eagle (Pl. 13)**
  Size medium to large; legs not feathered 16

9 Chest black or brown; legs not feathered to toes 10
  Chest dull pinkish; legs feathered to toes . . . .
    . . . . **Immature Crowned Eagle (Pl. 13)**

10 Chest and head black . . . . **Blackbreasted Snake Eagle (Pl. 13)**
    Chest brown; head pale; dark line through eye . . **Osprey (Pl. 15)**

11 Falcon head pattern (see diagram) 12
    No falcon head pattern 17

12 Falcon head pattern rufous; belly finely barred . . . .
    . . . . **Rednecked Falcon (Pl. 19)**
    Falcon head pattern blackish 13

13 Finely barred above and below . . . .
    . . . . **Immature Rednecked Falcon (Pl. 19)**
    Spotted and/or streaked below 14

14 Crown dark 15
    Crown pale or rufous 82

15 Robust build; inhabits cliffs and mountains . . . .
    . . . . **Peregrine Falcon (Pl. 19)**
    Slender build; inhabit open or wooded country . . . .
    . . . . . **Female Eastern Redfooted Kestrel, European Hobby
    Falcon, immature Eastern and Western Redfooted
    Kestrels (Pl. 19) (all very similar in the field)**

16 Sides of belly black; shape slender . . . .
    . . . . **Black Sparrowhawk (white-breasted phase) (Pl. 16)**
    Underparts all white; shape chunky; tail chestnut . . . .
    . . . . **Augur Buzzard (Pl. 15)**

17 Chest colour contrasts with belly colour 18
    Chest and belly colour the same 26

18 Chest grey; belly barred or streaked 19
    Chest black or brown 60

19 Belly streaked . . . . **Male Montagu's Harrier (Pl. 15)**
    Belly barred 20

20 Legs orange to reddish 21
    Legs yellow or paler 24

21 Rump boldly white 22
    Rump pale, but not pure white . . . .
    . . . . **Dark Chanting Goshawk (Pl. 16)**

22 Size larger; much white in wings; flight slow . . . .
    . . . . **Pale Chanting Goshawk (Pl. 16)**
    Size smaller; no large areas of white in wings 23

23 Vertical black line on throat; one broad white tailband . . . .
    . . . . **Lizard Buzzard (Pl. 15)**
    No black line on throat; four narrow white tailbands . . . .
    . . . . **Gabar Goshawk (Pl. 16)**

24 Face bare and bright yellow to reddish . . . . . **Gymnogene (Pl. 15)**
    Face feathered 25

25 Fine grey barring on belly; size relatively large; no rusty patch
       on nape 64
    Coarse rufous barring on belly; size medium; rusty patch on nape;
       slight crest . . . . **Cuckoo Hawk (Pl. 19)**

26 Ventral markings bold 27
    Ventral markings fine or pale 28

27 Size large 67
    Size small 72

28 Streaked or spotted 29
    Barred or obscurely mottled 30

29 Back plain rufous; head blue-grey . . . **Male Lesser Kestrel (Pl. 19)**
    Back buffy rufous like belly, boldly marked with black chevrons . .
    . . . . **Greater Kestrel (Pl. 19)**
    Back dark, unmarked 32

30   Rump white   31
     Rump not white   45

31   Barring grey; rump narrowly white . . . .
     . . . . **Ovambo Sparrowhawk (Pl. 16)**
     Barring rufous; rump more boldly white . . . .
     . . . . **Little Sparrowhawk (Pl. 16)**

32   Size large; throat dark; legs feathered to toes . . . .
     . . . . **Booted Eagle (Pl. 14)**
     Size small; throat pale; legs bare . . . .
     . . . . **Immature Little Sparrowhawk (Pl. 16)**

33   Belly black   34
     Belly brown, grey or rufous   35

34   Belly plain black   61
     Belly spotted or barred with white   73

35   Belly grey   36
     Belly brown or rufous   37

36   Undertail coverts rufous, contrasting with belly colour   78
     Undertail coverts grey like belly   79

37   Belly brown   43
     Belly rufous, or dark mahogany red   38

38   Size large; head, chest and tail white . . . .
     . . . . **African Fish Eagle (Pl. 13)**
     Head not white   39

39   Head blue-grey; chest marked with black; size small   71
     Head not blue-grey; crown dark or rufous   40

40   Falcon head pattern (see diagram)   83
     No falcon head pattern   41

41 Chest black; lower belly barred black and white; tail rufous . . . .
. . . . **Jackal Buzzard (Pl. 15)**
Chest rufous like belly 42
Chest white, streaked brown; head slender . . . .
. . . . **Immature Gymnogene (Pl. 15)**

42 Size very large; tail wedge-shaped; wings long . . . .
. . . . **Bearded Vulture (Pl. 10)**
Size small to medium; tail rounded 65

43 Whole bird appears dark uniform brown 44
Whole bird is dark or light brown and variously marked 53

44 Underwing white; crest long . . . . **Longcrested Eagle (Pl. 14)**
Underwing dark; crest short or absent 47

45 Back medium brown; belly mottled and/or barred with brown . . .
. . . . . . **Immature Pale and Dark Chanting Goshawks (Pl. 16)**
Back dark brown or slaty; belly finely barred 46

46 Tail narrowly banded (6–7 dark bands) . . . .
. . . . **Little Banded Goshawk (Pl. 16)**
Tail broadly banded (3–4 dark bands) . . . .
. . . . **African Goshawk (Pl. 16)**

47 Tail forked 48
Tail rounded or square 49

48 Bill yellow; head and body uniform . . . **Yellowbilled Kite (Pl. 15)**
Bill black; head paler than body . . . . **Black Kite (Pl. 15)**

49 Tail very short; head large; face bluish or orange . . . .
. . . . **Immature Bateleur (Pl. 13)**
Tail not very short 50

50 Pale throat contrasts with dark belly . . . . **Bat Hawk (Pl. 15)**
Throat and belly uniformly dark 51

51   Legs naked, whitish or yellow 76
     Legs feathered to toes, dark brown like belly 52

52   Pale windows in wingtips . . . . **Booted Eagle (dark phase) (Pl. 14)**
     No pale windows in wings . . . .
        . . . . **Wahlberg's Eagle (Pl. 14), Lesser Spotted Eagle,**
                Steppe Eagle, Tawny Eagle (dark phase) (Pl. 13)

53   Heavily marked below with mottling, spots or streaks of dark and
        light brown 54
     Not heavily marked below, or marked with light brown only   . . . .
        56

54   Heavily mottled or spotted below; pale spots on nape . . . .
        . . . . **Immature African Goshawk (Pl. 16)**
     Heavily streaked below 55

55   Shape slender; black line down centre of throat . . . .
        . . . . **Immature Black Sparrowhawk (Pl. 16)**
     Shape robust; ventral streaks broad 74

56   Shape slender; long wings held above horizontal in flight; sails low
        over ground 57
     Shape robust; broad wings held horizontal in flight 58

57   Leading edge of wing pale above; sometimes a pale band across dark
        chest; no white on rump . . . . **African Marsh Harrier (Pl. 15)**
     Leading edge of wing not markedly paler; general colour paler;
        rump distinctly white . . . .
        . . . . **Female Pallid and Montagu's Harriers (Pl. 15)**

58   Tail clearly barred or banded; legs partly or wholly naked and
        yellow 59
     Tail obscurely barred (barring not well visible in the field); legs
        feathered to toes; untidy tawny brown, sometimes irregularly
        mottled; some birds very pale . . . . **Tawny Eagle (Pl. 13)**

59 Two bands on tail; underparts look almost uniform dully streaked
brown; legs half feathered . . . . . . **Honey Buzzard (Pl. 15)**
Several fine bars on tail; underparts obscurely divided by paler
band on lower chest; often perched on poles 75

60 Chest plain unmarked brown 77
Chest streaky light brown; belly mottled brown and white;
obscure paler band on lower chest; legs bare; often perched
on telephone poles 75

61 Underwing white with black trailing edge; tail very short; mantle
brown; bare face and legs red . . . . **Bateleur (Pl. 13)**
Underwing dark; tail not very short 62

62 Pale windows in wingtips; size medium to large 63
No pale windows in wings; size small; tail faintly barred . . . .
. . . . **Gabar Goshawk (black phase) (Pl. 16)**

63 Size large; back white; legs feathered; mountain habitat . . . .
. . . . **Black Eagle (Pl. 13)**
Size medium; rump white; wings held above horizontal in flight;
usually in open country . . . . **Black Harrier (Pl. 15)**

64 Barring on belly distinct; underwing barred; two white bands on
tail . . . . **Southern Banded Snake Eagle (Pl. 14)**
Barring on belly faint; underwing white; one white bar on tail . . .
. . . . **Western Banded Snake Eagle (Pl. 14)**

65 Back barred; wings long 66
Back plain; wings short . . . . . **Redbreasted Sparrowhawk (Pl. 16)**

66 Belly boldly streaked black on light rufous . . . .
. . . . **Greater Kestrel (Pl. 19)**
Belly finely streaked brown on amber . . . .
. . . . **Female Western Redfooted Kestrel (Pl. 19)**

67 Streaked below 68
Spotted or barred below 69

68 Tail white with narrow black tip; underwings white; legs bare . . .
. . . . **Immature Fish Eagle (Pl. 13)**
Tail dark, banded; underwings dark; legs feathered to toes . . . . .
. . . . **African Hawk Eagle (Pl. 14)**

69 Spotted below; underwing blackish . . . . . . **Ayres' Eagle (Pl. 14)**
Barred below; underwing reddish . . . . . . **Crowned Eagle (Pl. 13)**

70 Falcon head pattern (see diagram); legs bare . . . .
. . . . **Lanner Falcon (Pl. 19)**
No falcon head pattern; legs feathered . . . . **Tawny Eagle (Pl. 13)**

71 Belly paler rufous; spotted (occasionally barred) with black; often
gregarious . . . . **Male Lesser Kestrel (Pl. 19)**
Belly darker rufous; streaked with black; usually solitary . . . . . .
. . . . **Rock Kestrel (Pl. 19)**

72 Buffy rufous all over; streaked below; bold black chevron markings
on back; underwing white . . . . **Greater Kestrel (Pl. 19)**
White below; back slaty; chest streaked; belly barred . . . .
. . . . **Immature Gabar Goshawk (Pl. 16)**

73 Breast chestnut; broad white wingbar in flight . . . .
. . . . **Jackal Buzzard (Pl. 15)**
Breast black, spotted with white . . . .
. . . . **Black Sparrowhawk (black-bellied phase) (Pl. 16)**

74 Legs bare, yellow . . . . **Immature Steppe Buzzard (Pl. 15)**
Legs feathered to toes . . . . **Immature Tawny Eagle (Pl. 13)**

75 Belly mainly barred . . . . **Steppe Buzzard (Pl. 15)**
Belly mainly streaked . . . . **Forest Buzzard (Pl. 15)**
(These two species are very variable in plumage and are similar
in appearance, therefore hard to tell apart in the field)

76 Legs yellow; bill yellow; head slender . . . .
. . . . **Immature Gymnogene (Pl. 15)**

Legs whitish; bill black; head large; eyes yellow and owl-like   . . .
. . . . **Brown Snake Eagle, Immature**
Blackbreasted Snake Eagle (Pl. 14)

77 Chest dark brown; belly spotted; size large; legs feathered to
toes . . . . **Martial Eagle (Pl. 15)**
Chest light brown; belly barred; size medium; legs bare . . . .
. . . . **Immature Dark Chanting Goshawk (Pl. 16)**

78 Underwing white . . . . . **Male Eastern Redfooted Kestrel (Pl. 19)**
Underwing dark . . . . . . **Male Western Redfooted Kestrel (Pl. 19)**

79 Whole body appears uniform grey 80
Body grey; head paler (whitish) . . . . **Dickinson's Kestrel (Pl. 19)**

80 Primaries dark; wings extend beyond tail when perched; tail not
barred; dark mark under eye . . . . **Sooty Falcon (Pl. 19)**
Primaries and tail lightly barred; wings much shorter than tail
when perched . . . . **Grey Kestrel (Pl. 19)**

81 Body all grey; head whitish; legs yellow . . . .
. . . . **Dickinson's Kestrel (Pl. 19)**
Undertail coverts rufous; legs orange . . . .
. . . . **Male Eastern Redfooted Kestrel (Pl. 19)**

82 Back dark slate grey or brown . . . . **Lanner Falcon (Pl. 19)**
Back rufous, barred with black . . . **Female Lesser Kestrel (Pl. 19)**

83 Streaked below 84
Not streaked below . . . . **African Hobby Falcon (Pl. 19)**

84 Crown all dark . . . . **Immature African Hobby Falcon (Pl. 19)**
Crown rufous . . . . . **Immature Western Redfooted Kestrel (Pl. 19)**

(*Roberts'* pages 429–452; Plates 44–45)

1 Belly rufous, buff or white 6
   Belly black, or with black central patch 2

2 Conspicuous white ear patches 3
   No conspicuous white ear patches 4

3 Back grey or sandy . . . . **Male Greybacked Finchlark**
   Back rufous . . . . **Male Chestnutbacked Finchlark**

4 Back grey or sandy . . . . **Female Greybacked Finchlark**
   Back rufous 5

5 Head and belly all black; no pale collar on hindneck . . . .
   . . . . **Male Blackeared Finchlark**
   Chest mottled greyish; black patch in centre of belly; pale collar
   on hindneck . . . . **Female Chestnutbacked Finchlark**

6 Belly white 7
   Belly rufous or buff 18

7 Crown and pectoral patches plain bright rufous . . . .
   . . . . **Redcapped Lark**
   Crown mottled, streaked or dark brown; no pectoral patches . . . .
   8

8 Chest streaked or spotted 9
   Chest plain white; back pale pinkish . . . . **Gray's Lark**

9 Tail very short and narrow (appears tailless) . . . . **Rudd's Lark**
   Tail not noticeably short and narrow 10

10 Crown and back plain dark brown; breast and head boldly marked
   dark brown on white . . . . **Dusky Lark**
   Crown and back light brown or rufous, streaked or spotted . . 11

11 Bill rather heavy with yellow base . . . . **Thickbilled Lark**
   No yellow at base of bill 12

12 Clear white eyebrow 13
   Eyebrow absent or indistinct; whitish ring around eye 27

13 Bill nearly as long as head, decurved . . . . **Longbilled Lark**
   Bill shorter than head, rather stout 14

14 White above and below eye 16
   White above eye only 15

15 White eyebrow does not extend to base of bill; primaries edged
     rufous; ear coverts heavily streaked . . . . . **Monotonous Lark**
   White eyebrow extends to base of bill; primaries edged grey;
     ear coverts lightly streaked . . . . **Sabota Lark**

16 Dark line from gape to ear coverts 17
   No dark line from gape to ear coverts . . . . . . **Fawncoloured Lark**

17 Bill rather slender, paler at base; habitat not confined to red
     sand on Bushman Flats 28
   Bill rather stout and uniformly dark; confined to red sand on
     Bushman Flats . . . . **Red Lark**

18 Bill pink 19
   Bill not pink 20

19 Belly uniform light rufous; flanks not streaked . . . **Pinkbilled Lark**
   Belly paler towards tail; flanks streaked blackish; confined to
     upper Vaal River catchment area . . . . **Botha's Lark**

20 Tail very short and narrow (appears tailless) . . . . **Rudd's Lark**
   Tail not noticeably short and narrow 21

21 Bill rather slender and decurved 22
   Bill rather stout and conical 23

22 Throat pure white; eyebrow rufous or buff; tail tipped white . . .
. . . . Spikeheeled Lark
Throat dull white; eyebrow white; tail all dark . . . .
. . . . Shortclawed Lark

23 Dark almost vertical mark below eye; no rufous in wing . . . .
. . . . Sclater's Lark
No vertical mark below eye; primaries edged rufous 24

24 Rattling or clapping flight display; back obscurely mottled . . . 25
No rattling flight display; back fairly distinctly streaked 26

25 Rattling display flight followed by drawn-out whistle . . . .
. . . . Clapper Lark
Rattling display flight followed by chippering song . . . .
. . . . Flappet Lark

26 Rather large, chunky bird; bill heavy and longish . . . .
. . . . Rufousnaped Lark
Smallish bird; bill fairly short; throat clear white . . . .
. . . . Melodious Lark

27 Back rufous; head does not appear crested . . . .
. . . . Female Blackeared Finchlark
Back sandy buff; head often appears crested . . . . . . Stark's Lark

28 Back greyish or rufous, boldly streaked . . . . Karoo Lark
Back rufous to pinkish, faintly streaked or nearly plain . . . .
. . . . Dune Lark

(*Roberts'* pages 576–595; Plates 55–56)

1 Tail proportionately shorter (24–54 mm); habitat mostly open
   grassland, flat or gently sloping, less often with trees   2
   Tail proportionately longer (45–72 mm); habitat rank grassland,
   vlei or marsh, or grassy understory of woodland or savanna
     8

2 Back grey, unmarked; tail plain (without bold pattern at tip);
   habitat with bushes and trees   3
   Back buff, streaked blackish; tail boldly patterned at tip; habitat
   open grassland without trees   4

3 Crown plain grey or dull rufous, streaked blackish in winter; below
   whitish; tail very short (29–35 mm) in summer; song feeble
   see see see . . . . **Shortwinged Cisticola**
   Crown plain dull rufous; tail longer than 37 mm; below grey;
   song penetrating *weep weep weep* . . . . **Neddicky**

4 Back lightly streaked; rump buff or greyish; song high-pitched
   ting-ting-ting, one note/half second . . . . . . **Desert Cisticola**
   Back boldly streaked; rump bright tawny rufous   5

5 Display flight low (up to about 10 m); song monotonous unmusical
   zit zit zit, one note/second . . . . **Fantailed Cisticola**
   Display flight very high (bird usually out of sight); song of two
   different notes in set phrase   6

6 Song quick *see see see chik chik chik* . . . . **Cloud Cisticola**
   Song slower, more deliberate   7

7 Song short (fewer than 7 notes), repeated frequently over long
   period of time . . . . **Ayres' Cisticola**
   Song long (10–20 or more notes), repeated infrequently . . . .
   . . . . **Palecrowned Cisticola**

8   Crown plain rufous, or only lightly streaked   9
Crown grey, streaked darker; tail grey, boldly patterned with black
and white at tip; bill heavy, decurved; voice loud croaking . .
. . . . **Croaking Cisticola**

9   Back grey, streaked darker, or back black, scaled with buff or
grey 12
Back plain grey (washed rusty in winter) 10

10   Tail without bold pattern at tip, often held cocked upwards; call
notes loud and petulant; song jumbled phrases . . . .
. . . . **Lazy Cisticola**
Tail boldly patterned at tip 11

11   Crown rufous, contrasting with brown back; tail greyish brown;
habitat bracken-briar or streamside bushes and trees, mostly
above 1000 m . . . . **Singing Cisticola**
Crown and back similarly rufous; tail russet; habitat rank marshy
vegetation, mostly at lower altitudes . . . . **Redfaced Cisticola**

12   Habitat marshy 13
Habitat not marshy 15

13   Back black, scaled buff or grey 14
Back grey, streaked black; tail dusky brown, looking broad in
flight; below rusty buff; song twanging and croaking phrase
of 4 notes; northern Botswana only . . . . . **Chirping Cisticola**

14   Rump dull olive-buff, streaked black; tail rusty brown; distribution
mainly highveld and extreme south . . . **Levaillant's Cisticola**
Rump plain grey; tail grey, tipped black; distribution mainly low-
lying e regions and n Botswana . . . . . . **Blackbacked Cisticola**

15   Habitat usually open without bushes, or with low shrubs
only 16
Habitat with bushes and trees 17

16 Breast greyish white; back lightly streaked; habitat semi-arid; distribution mainly western . . . . **Greybacked Cisticola**
Breast buff; back boldly streaked; habitat mostly montane, steep and often rocky; mainly eastern distribution . . . .
. . . . **Wailing Cisticola**

17 Tail bright rufous; lores and eyebrow buff; habitat scrub or secondary growth; song bell-like *tweee tweee tweee*; behaviour secretive . . . . **Tinkling Cisticola**
Tail dusky brown; lores and eyebrow greyish white; habitat *Acacia* savanna, woodland, coastal scrub; song set phrase of 2–4 introductory notes followed by trill or rattle; behaviour bold . . . . **Rattling Cisticola**

(*Roberts'* pages 730–742; Plates 67–68)

1 Bill red or yellow . . . . **Redbilled Quelea adult**
  Bill not red or yellow 2

2 Throat and breast plain, faintly mottled or faintly streaked . . 3
  Throat and breast clearly streaked 7

3 Wrist patch yellow . . . . **Male Yellowbacked Widow**
  Wrist patch not yellow 4

4 Crown boldly streaked 5
  Crown faintly streaked 6

5 Breast faintly streaked . . . . **Female Yellowbacked Widow**
  Breast plain, pale tawny . . . . **Redcollared Widow**

6 Crown greyish brown . . . . **Immature Redbilled Quelea**
  Crown yellowish brown . . . . **Redheaded Quelea**

7 Wrist patch same colour as rest of body; tail short (less than one
    quarter length of body) 8
  Wrist patch rusty or yellowish; tail longish (more than one quarter
    length of body) 10

8 Build robust; back dull olive-brown; belly pale buff 9
  Build slender; back dull rufous-brown; belly almost white . . . . . .
    . . . . **Golden Bishop**

9 Wings and tail brown . . . . **Red Bishop**
  Wings and tail black . . . . **Firecrowned Bishop**

---

*Based on a key by Johnson, D. N. & Horner, R. F. (1986, *Bokmakierie* 38:13–17),
with modifications.

10 Wrist patch rusty 11
   Wrist patch yellow 12

11 Size large (length more than 18 cm); whole breast streaked . . . . .
   . . . . **Longtailed Widow**
   Size smallish (length less than 18 cm); breast streaked at sides
   only . . . . **Redshouldered Widow**

12 Wrist patch dull yellow; rump brown . . . . . . **Whitewinged Widow**
   Wrist patch olive-yellow; rump yellow . . . **Yellowrumped Widow**

THE SMALL CORMORANTS (*Roberts'* Plate 4)

1 Tail relatively short; eye green or brown; marine 2
  Tail relatively long; eye red 3

2 Face yellow . . . . **Cape Cormorant**
  Face black . . . . **Bank Cormorant**

3 Freshwater . . . . **Reed Cormorant**
  Marine . . . . **Crowned Cormorant**

THE LARGE GREY HERONS (*Roberts'* Plate 5)

Bill yellow; neck white; black line through eye . . . . **Grey Heron**
Bill blackish; neck and crown black or slaty . . . .
. . . . **Blackheaded Heron**

THE WHITE EGRETS (*Roberts'* Plate 5)

1 Legs and feet all black; bill black or yellow; black line extending
  to behind and below eye . . . . **Great White Egret**
  Legs or feet with touch of yellow; line under eye faint or
  absent 2

2 Bill black; legs black; toes yellow . . . . **Little Egret**
  Bill yellow or orange 3

3 Legs and toes black; thighs yellowish; bill yellow; neck long . . . .
  . . . . **Yellowbilled Egret**
  Legs dull yellow or greenish; tinge of pinkish-buff on head, back
  and chest; neck shortish . . . . **Cattle Egret**

THE BLACK STORKS (*Roberts'* Plate 7)

1 Neck white . . . . **Woollynecked Stork**
  Neck black 2

2 Belly black . . . . **Openbilled Stork**
  Belly white 3

3 Bill and legs all red . . . . **Black Stork**
  Bill black; legs black, except for red toes and 'knees'; face blue . .
  . . . . **Abdim's Stork**

THE WHITE STORKS (*Roberts'* Plate 7)

  Bill red, pointed; tail white . . . . **White Stork**
  Bill yellow, curved at tip; tail black . . . . **Yellowbilled Stork**

FLAMINGOES (*Roberts'* Plate 7)

  Body appears pure white at distance; bill very pale with black
    tip . . . . **Greater Flamingo**
  Body appears pink at distance; bill very dark . . . **Lesser Flamingo**

THE LARGE BUSTARDS (*Roberts'* Plate 24)

1 Hindneck rufous; neck not barred 2
  Hindneck not rufous; neck barred . . . . **Kori Bustard**

2 Crown black; wings black-and-white; foreneck grey . . . .
  . . . . **Stanley's Bustard**
  Crown brown; wings brown-and-white; foreneck brown . . . .
  . . . . **Ludwig's Bustard**

DIKKOPS   (*Roberts'* Plate 25)

Back streaked; pale bar on folded wing . . . . **Water Dikkop**
Back spotted; no pale bar on folded wing . . . . **Cape Dikkop**

UNCOLLARED PLOVERS   (*Roberts'* Plate 26)

Black line behind white front does not meet eye . . . .
. . . . **Whitefronted Plover**
Black line behind white front meets eye; broad black stripe from
behind eye to neck . . . . **Kittlitz's Plover**

BLACK-WINGED PLOVERS   (*Roberts'* Plate 26)

White of front extends to above eye; secondaries tipped with black;
eyering scarlet; legs dull red . . . . **Blackwinged Plover**
White of front stops in front of eye; secondaries completely white;
eyering brown; legs brown . . . . . . **Lesser Blackwinged Plover**

PLAIN RUFOUS COURSERS   (*Roberts'* Plate 30)

Black chestband extends onto belly and between legs; hindcrown
rufous . . . . **Temminck's Courser**
Black chestband does not extend onto belly; hindcrown blue . . . .
. . . . **Burchell's Courser**

THE WHITE-RUMPED SWIFTS   (*Roberts'* Plate 39)

1 Tail square; rump broadly white . . . . **Little Swift**
   Tail forked 2

2 Tail deeply forked; white on rump narrow, V-shaped . . . .
   . . . . **Whiterumped Swift**
   Tail slightly forked; white on rump broad and straight . . . .
   . . . . **Horus Swift**

THE SMALL, BLUE, RED-BILLED KINGFISHERS
(*Roberts'* Plate 40)

Crown turquoise blue; no chestnut eyebrow; waterside habitat . .
. . . . **Malachite Kingfisher**
Crown violet blue; chestnut eyebrow; woodland habitat . . . .
. . . . **Pygmy Kingfisher**

THE YELLOW-VENTED, CRESTED BULBULS (*Roberts'* Plate 48)

1 No eyering visible (eyering black) . . . . **Blackeyed Bulbul**
   Conspicuous coloured eyering 2

2 Eyering white . . . . **Cape Bulbul**
   Eyering red . . . . **Redeyed Bulbul**

THE GREEN-AND-YELLOW SHRIKES (*Roberts'* Plate 61)

1 Black collar on chest 2
   No black collar on chest 3

2 Throat red; head green . . . . **Gorgeous Bush Shrike**
   Throat yellow; head grey . . . . **Bokmakierie**

3 Pinkish white below . . . . **Olive Bush Shrike (ruddy phase)**
   Orange and/or yellow below 4

4 Head green . . . . **Olive Bush Shrike (olive phase)**
   Head grey 5

5 Face grey like crown; bill very heavy . . . **Greyheaded Bush Shrike**
   Face black 6

6 Yellow eyebrow separates black face from grey crown . . . .
   . . . . **Orangebreasted Bush Shrike**
   No yellow eyebrow . . . . **Blackfronted Bush Shrike**

THE TCHAGRAS (*Roberts'* Plate 61)

1 Pale eyebrow present 2
  No pale eyebrow . . . . Marsh Tchagra

2 Crown black . . . . Blackcrowned Tchagra
  Crown brown or rusty 3

3 Crown rusty; no black line above pale eyebrow; belly white . . . .
  . . . . Southern Tchagra
  Crown brown; thin black line above pale eyebrow; belly buff . . .
  . . . . Threestreaked Tchagra

THE YELLOW WEAVERS (ADULT MALES IN BREEDING PLUMAGE) (*Roberts'* Plate 66)

1 Back and crown plain black . . . . Forest Weaver
  Back greenish or mottled 2

2 Face mask black 3
  Face mask brown, green or absent 6

3 Black of face mask extends to front and/or crown 4
  Black of face mask does not extend to front (front orange or
  yellow) 5

4 Mask extends in narrow point onto chest; iris red 10
  Mask rounded below; crown mostly black; iris cream . . . .
  . . . . Lesser Masked Weaver

5 Mask broad; crown yellow; back spotted . . . .
  . . . . Spottedbacked Weaver (*spilonotus*)
  Mask forms narrow bib only; crown orange-yellow; back plain
  green . . . . Spectacled Weaver

6 Face mask distinct, brown or green 7
  Face mask indistinct or absent 8

7 Face mask brown . . . . **Brownthroated Weaver**
   Face mask green; chest brown . . . . **Oliveheaded Weaver**

8 Face rich chestnut-orange grading into yellow on body and head;
   iris cream . . . . **Cape Weaver**
   No chestnut on head; iris red or yellow 9

9 Canary yellow below; iris red . . . . **Yellow Weaver**
   Golden yellow below; iris yellow . . . . **Golden Weaver**

10 Crown yellow; back faintly streaked greenish . . . **Masked Weaver**
    Crown black; back heavily mottled . . . .
    . . . . **Spottedbacked Weaver (*nigriceps*)**

THE YELLOW CANARIES (MALES ONLY) (*Roberts'* Plate 72)

1 Back green 3
   Back brown 2

2 Tail and wings tipped white . . . . **Cape Siskin**
   Tail and wings not tipped white . . . . **Drakensberg Siskin**

3 Nape grey . . . . **Cape Canary**
   Nape green like back 4

4 Chin black; belly streaked with green . . . . **Forest Canary**
   Chin yellow; belly not streaked 5

5 Rich sulphur yellow below, washed greenish on breast; bill very
   heavy . . . . **Bully Canary**
   Clear yellow below; bill not very heavy 6

6 Bright deep yellow below, including flanks; rump dull greenish
   yellow; western distribution; female dull streaky grey . . . . .
   . . . . **Yellow Canary**
   Light yellow below, fading to greyish on flanks; rump bright
   yellow; eastern distribution; sexes alike . . . .
   . . . . **Yelloweyed Canary**

SOME USEFUL BOOKS ON SOUTHERN AFRICAN BIRDS

Berruti, A. & Sinclair, J. C. 1983. *Where to watch birds in southern Africa.*
Cape Town: Struik.

Bruun, B. & Singer, A. 1970. *The Hamlyn guide to the birds of Britain
and Europe.* London: Hamlyn.

Clancey, P. A. 1964. *The birds of Natal and Zululand.* Edinburgh: Oliver
& Boyd.

Clancey, P. A. 1971. *A handlist of the birds of southern Moçambique.*
Lourenço Marques: Investigaçao Cientifica de Moçambique.

Cyrus, D. P. & Robson, N. F. 1980. *Bird atlas of Natal.* Pietermaritzburg: University of Natal Press.

Gill, E. L. 1968. *A first guide to South African birds.* Cape Town:
Maskew Miller. (Revised by J. M. Winterbottom)

Ginn, P. 1973. *Birds afield.* Rhodesia: Longman. (Bundu Series)

Harrison, P. 1983. *Seabirds: an identification guide.* Beckenham: Croom
Helm.

Hayman, P., Marchant, J. & Prater, T. 1986. *Shorebirds: an identification
guide to the waders of the world.* Beckenham: Croom Helm.

Heinzel, H., Fitter, R. & Parslow, J. 1972. *The birds of Britain and
Europe, with North Africa and the Middle East.* London: Collins.

Irwin, M. P. S. 1981. *The birds of Zimbabwe.* Salisbury: Quest.

Mackworth-Praed, C. W. & Grant, C. H. B. 1962–1963. *Birds of the
southern third of Africa.* Vols 1 and 2. London: Longmans.

Maclean, G. L. 1965. *Roberts' birds of southern Africa.* Cape Town:
John Voelcker Bird Book Fund.

McLachlan, G. R. & Liversidge, R. 1978. *Roberts' birds of South Africa.*
Johannesburg: John Voelcker Bird Book Fund. (Fourth edition)